S: To my eternal Bride and soul mate, your faith saved me more times than I can count. I look forward to always walking hand in hand with you through good times and challenges.

V & curlyV: To my amazing daughters, your intelligence, beauty and emotional strength makes me so proud. You keep me focused and give me purpose.

S, V, curlyV: You girls have been so supportive and understanding when I went to my desk or some other place in the world to study and research, all the while never complaining. Upon my return, you always have a beautiful smile and a hug ready for me.

This book and all the other "stuff" I have done to help protect others would not be possible without your incredible support.

Thank you and I Love you,

Mark/Daddy

Copyright-All Rights Reserved by Mark Carey

Copyright-All Rights Reserved by Mark Carey

Table of Contents

Abstract..1

Executive Summary...2

System Architecture
    Core Environment......................................5
    Neural Network Core..................................5
    Intelligence Core...6
    Site Activity Core...7
        Data Lake for Onsite Images............7
        Alerting Environment..........................7

Site Environment
    Camera Surveillance..................................9
    AI-FRTD Collector....................................10
    Network Aggregator................................11

Monitoring..12

Technology Background
    Artificial Intelligence Conceptual View..........13
    The Neuron...16
    Artificial Neural Network........................19
    Intelligence...20
    Computer Vision and Facial Recognition......24
    Convolutional Neural Network
            for Facial Recognition............27

Threat Assessment..29

Social Media..31

Feature Add-On..33

Acronyms..34

Appendix..36

Bibliography..44

Copyright-All Rights Reserved by Mark Carey

Copyright-All Rights Reserved by Mark Carey

ABSTRACT

The five deadliest shootings in U.S. history have occurred in the last 11 years. The targets are innocent people ranging from very young children in schools to adults in the workplace and at social gatherings. Although the number of incidents has fluctuated each year, overall there is an increasing trend in both active shooter incidents and the casualties they cause. We must accept as a society that we are not doing enough to stop this violence and consider all possibilities to protect our families.

This book will not take a position on gun policy, rather it is intended to offer one of several technical architectures that will enhance security in our homeland by providing an early warning detection of a potential mass shooter onsite in either a school, workplace, government facility or a place of gathering.

No matter when American society comes to an agreement on gun policy, many guns will exist for a long time in the United States. Therefore, along with the immediate need to detect and prevent active shooters, there will continue to be a need for a solution to protect innocent people against a mass shooter.

The intent of this work is not to detail the components of a solution, but rather to provide a realistic Architectural design of a protective solution, that utilizes the power of Artificial Intelligence Facial Recognition.

EXECUTIVE SUMMARY

The five deadliest shootings in United States history have occurred in the last 11 years. (*Appendix A*). The targets were innocent people ranging from very young children in schools to adults in the workplace and at social gatherings. In fact, according to the FBI, over 64 percent of Active Shooter incidents from 2000 to 2016 have occurred in education centers and places of work. (*Appendix B*). Although the number of incidents has fluctuated each year, overall there is an increasing trend in both active shooter incidents (*Appendix C*) and their casualties (*Appendix D*).

During the first 5 months of 2018, there have been over 22 school shootings that have resulted in injury or death (Ahmed, 2018). Just two of this year's school shootings, one at Florida's Marjory Stoneman Douglas High School and the other at Texas' Santa Fe High School have together claimed a total of 27 innocent lives (Coaston, et al.,2018). We must accept as a society that we are not doing enough to stop this violence and consider all possibilities to protect our children and loved ones. In other words, "Enough is Enough".

There are those who call for stricter gun laws in the hopes of curbing the rise of gun violence. Conversely there are those that fear tightening of gun ownership policy is a veiled attempt to deny what is interpreted as a fundamental right to gun ownership granted by the second amendment. The clause Constitutional advocates point to reads: "A well regulated Militia, being necessary to the security of a free State, the right of the people to keep and bear Arms, shall not be infringed." (Cornell Law School, n.d.)

It should be noted that gun policy is not a panacea for mass shootings, in fact no single solution will solve this crisis. Policy can only be a compliment to other solutions that work to resolve the many root causes of gun violence. Deeper societal issues cannot be solved with regulations and policies, like mental health issues, social ills (Knoll-Annas,2016), terrorist motivations (Crenshaw, 2017), and extremism (Williams, 2017).

This paper will not offer a position on the gun policy debate, rather it is intended to offer one of several technical architectures that will enhance security in our homeland by eliminating mass shootings in schools, workplaces, government facilities and where our citizens gather. No matter when American society comes to an agreement on gun policy, many guns will always exist in the United States. Therefore, along with the immediate need to detect and prevent active shooters, there will continue to be a need for a solution to protect innocent people against those with a desire to harm others and disregard for legal gun policy.

Privacy advocates serve a very important function in American society. They defend the right to privacy, protect the innocent from subjective judgement, prevent fear of those we grant power to keep us safe, prevent abuse of power and ultimately protect society from degradation due to the population's mental anguish of paranoia. However, we must temper this incredibly important right with a pragmatic and open-minded view that that balances the priority of privacy and physical security.

It is doubtful the intent of our founding fathers and those that protect the right to privacy, feel privacy is more important than life itself. To help qualify the priority of privacy as it relates to the objective of reducing mass shootings, the following question should be considered:

"When entering an area where children or adults gather, what is more important to an innocent person or their child:

1) Protecting an innocent person from being identified in police databases or what they have said in social media sites?
-or-
2) Protecting them from death or severe injury?"

As privacy advocates must consider the above question of privacy over safety, so too should Security Architects-and those they serve-step back and look at their design to ensure the right to privacy is not being ignored.

Therefore, a good security architecture that protects innocent lives against physical harm should also protect the right to privacy as much as possible, since protecting privacy also protects innocent lives.

Identity recognition has evolved far past Galton's 1892 finger print classification (FindLaw, 2018) system. With the growing use of Artificial Intelligence, facial recognition has evolved to the point that identification of an individual can be done quickly, with a high degree of accuracy, and without requiring direct contact. It is not yet, as reliable as identifying a fingerprint, but just like fingerprint technology has done over the last 120 years, the solution will evolve and continue to be a valuable assistive technology for law enforcement to protect and serve.

To implement the architecture proposed herein, nothing must be invented, rather the components must be assembled to meet the objective. Since this paper is limited in length, explanations of the main technologies required to build the solution are not presented in exhaustive detail. Technical summaries of certain components that are required are presented to validate the architecture and provide an overall path to the implementation of the technology. In addition, only the visual portion (camera) of the protection solution and its associated Convolutional Neural Network is presented. Additional components of the solution that can provide no-touch weapons and body armor detection are outside the scope of this paper and will be offered in subsequent research presentation.

This work is presented with the hopes of better protecting children and adult citizens against those who threaten them with shootings, bombings and kidnappings. The following proposed architecture is built upon Artificial Intelligence and current computer facial recognition technology.

SYSTEM ARCHITECTURE

The following presents the major components of the architecture diagram contained in: *Appendix E: Artificial Intelligence Facial Recognition Threat Detection Architecture Diagram.*

CORE ENVIRONMENT

Neural Network Core

It is important to first clarify that an Artificial Intelligence based Facial Recognition Threat Detection (*AI-FRTD*) system is a provider of intelligence and may only send alerts for individuals of concern to appropriate law enforcement or security personnel. The AI-FRTD system must not be granted authority to command or send recommendations on how to proceed. Law enforcement personnel and those sensitive to the security needs of a site are best equipped to make the decision on how to best proceed.

The Neural Network Core of the AI-FRTD is in a highly secure Hyperscale Data Center (Recap, n.d.) with a high degree of redundancy, including geographic dispersion. It will receive images from the various cameras located onsite at the protected remote areas. The AI-FRTD Neural Network Core system will analyze the images using Convolutional Neural Network (ConvNet or CNN) techniques to isolate faces from an image then identify and catalog their uniqueness.

Intelligence Core

The results of the ConvNet analysis of the camera images are compared against ConvNet facial identity searches of the images contained in the 14 National Crime Information Center (NCIC) persons databases (FBI-NCIC, n.d.) and the DHS Customs and Border Protection facial biometric exit system (DHS-CBP, 2017) database. Matches will be sought for those in the NCIC database of people that require contact by a law enforcement, such as outstanding warrants for arrest, previous convictions for weapons violation, or convictions for sexual offenses (especially if the images are received from school areas and areas where children gather). Also, ConvNet facial image recognition will be conducted of the DHS-CBP exit database for individuals who recently departed to a country listed in the State Sponsors of Terrorism (DOJ, n.d.).

For matches made in the federal databases, the pertinent information on the individual is retrieved from the database for input into an alert notification file that contains the name of the individual, their image sent by the onsite camera, and the Global Positioning System (GPS) coordinates for the camera that captured their image. The alert file is sent to the responsible security command for the location, in addition the same alert file should be sent to the local law enforcement agency. The command or law enforcement agency can then forward this alert file to the individual(s) dispatched to intercept the subject and other personnel that should be notified.

In addition to examining federal databases, the ConvNet analysis of the camera images will also be used to conduct Social Media Intelligence (SOCMINT) research by comparing the users' facial images in multiple social media sites such as Facebook, Twitter, and LinkedIn. If an image match is made in a social media site, their social media files will be searched for key words to detect warnings of impending attacks (leakage).

Site Activity Core

Data Lake for Onsite Images

After the AI-FRTD Collector completes its image capture and adds the appropriate metadata to the image, it will be forwarded to the Onsite Images Data Lake in the Core Environment. The Data Lake is a database with large data storage capacity with minimal processing done to the received images. A Data Lake model will be best to store the images since the ConvNet of the AI-FRTD Neural Network Core will be applying its algorithms which are optimal for image processing.

Alerting Environment

For leakage matches made in social media sites, an alert file is created that contains the name of the individual, their image sent by the onsite camera, the GPS coordinates of the camera that captured their image and their threat (leakage statement). The alert file is sent to the person or group responsible for security of the location. This alert file can be forwarded to the individual(s) dispatched to intercept the subject and other personnel that should be notified.

Two points regarding alerts should be noted. First, although both the federal database and SOCMINT searches both require an alert to be sent, there are key differences between them. Specifically, alerts from a hit on a federal database requires law enforcement to be notified, whereas a SOCMINT alert requires it to be sent to those responsible for the security of the monitored location. The SOCMINT communication process could be modified to also include law enforcement in initial or subsequent alerts. Required notifications for a monitored site is customizable in the Alerting Environment. Second, alerts received by a recipient can be forwarded by the recipient to another responsible party, for example the school administrator responsible for security could forward an alert they received to local law enforcement personnel or dispatch desk.

To provide an effective alert, an alert application must be created that will be compatible with Smartphones and cellular connected tablets running either the Android or IOS operating system. The application will receive alerts and provide the capability to respond to the alert. If a response to an alert is made, it will be sent back to the Alerting Environment. The response will include the responder identity and the date and time of their response. If a response is not received from one of the alert recipients within a predetermined time, then subsequent alerts will be sent that conforms to a predetermined escalation notification plan established in the Alerting Environment for the specific site.

Both the cellular infrastructure and Internet could be used to communicate alerts and receive their confirmation; however, some notification controls would have to be incorporated in the alert application in the Alerting Environment to eliminate redundant alerts being sent.

Since the alerts contain GPS coordinates, the application should provide integration with the smartphone or tablet mapping application. This will help direct the responder to the location of the camera that captured the individual's image.

SITE ENVIRONMENT

Camera Surveillance

Both still and video cameras have become very inexpensive. Cameras are available in all sizes from cameras the size of pinholes to those requiring solid mounting. In addition to their lower cost and size options, newer cameras provide high resolution image quality, many can see in near total darkness, many can connect back to video recorders wirelessly, some require only low power, there are wearable types, they are built into drones that can be driven remotely while sending back their images or video feeds over a wireless network back to storage devices. The reality is that video and still cameras have evolved to the point of being very flexible, affordable and highly connected.

For an area that requires monitoring for safety, like a school, a workplace or large public gathering location, cameras can be mounted in areas that monitor ingress points and in locations around the area to be monitored. Wearable cameras such as vest mounted cameras or glasses could be deployed on security personnel, law enforcement personnel, or canines, if mobile safety monitoring is required. In addition to the statically mounted and human mounted cameras, ones can be mounted on drones to act in a sentry manner and be deployed on demand to an area of interest or elevated risk such as an active shooter situation.

AI-FRTD Collector

Both stationary cameras and mobile cameras send their still images or video streams to an onsite image collector that performs additional processing before sending it to the Core Environment for final analysis.

The onsite image collector is referred to as the Artificial Intelligence Facial Recognition Threat Detection Collector or AI-FRTD Collector. The *AI-FRTD Collector* will be connected to the LAN via Ethernet and will accept images sent to it by the onsite cameras. The AI-FRTD Collector will accept still images from the onsite still image cameras and will capture a still image of faces from the video feeds it receives. It will then add metadata values to the still images before sending to the Onsite Images Database in the Site Activity environment. The metadata values added by the AI-FRTD Collector will include an identifier of the site, date and time of the image received from a camera and a camera identifier.

Communications between AI-FRTD Collector and the Site Activity environment will require an encrypted Internet connection with an applied appropriate bandwidth priority. Another connection type between the two environments could be a dedicated Telecommunications connection. To protect the systems from other systems, Security Gateways with very strict access control rules and segmentation between the Internet, LAN and AI-FRTD Collector will be configured. The Security Gateways also monitor, and alert on malicious activity and service disruption attempts such as Denial of Services (DoS) attacks.

Encrypted Internet connectivity may be used as a connection type to send images from the sites or receive them at the Onsite Image Data Lake in the Core Environment. In addition, the Alerting Environment could also send alerts via the Internet. Solely relying on a single network connectivity type increases the risk exposure, such as an interference to the site location's Internet connection. Specifically, if a monitored site's Internet connection is interrupted images may not reach the Onsite Images Database and alerts may not be received. Therefore, redundant network connections should be provided for communication between the Site Environments and the Core Environment.

Along with offering encrypted Internet connectivity, dedicated Telecommunications Leased circuits and the existing U.S. cellular infrastructure are also recommended as a communication system. If using the U.S. cellular infrastructure, systems should be registered with the Government Emergency Telecommunications System (GETS) (DHS-GETS, 2017). Devices using cellular that are registered on the GETS system receive priority for their connection over other non-registered cellular devices. Having priority during times of high traffic volumes will help improve the chance to send images, such as during an attempted Active Shooter situation. The negative impact of a saturated local cellular infrastructure was evidence during the 2013 Boston Marathon bombings when, due to a high volume of calls after the bombings, cellular phone service became unreliable for emergency response personnel (Ungerleider, 2013)

Network Aggregator

At a site, there is a Network Aggregator that centralizes both the wired Local Area Network (LAN) switches and the Wireless Access Points (WAPs) to ensure surveillance cameras can connect with the AI-FRTD Collector.

Wireless Access Point (WAPs) will be connected to the LAN and disturbed in such a manner to ensure wireless network availability to the cameras that cannot connect directly to the LAN. The wireless coverage is especially important for Mobile Cameras.

MONITORING

Due to the criticality of such a system, oversight of all parts of the environment is required. This will be done with a separate environment referred to as the AI-FRTD Environment Status Monitoring system.

The environment presents a management view for each AI-FRTD site. The management view should include the status of the AI-FRTD Security Gateway network devices, AI-FRTD Collector, cameras, alerts and responses to the alerts. Additionally, the management system should provide a mechanism to allow administrators of specific AI-FRTD systems to enter notes and manual logs to be reviewed by other administrators. Access to the management interfaces is controlled by Role Based Access Control (RBAC) authentication and authorization, which is based on the individual's authority for their specific site. In addition to site specific information, authority and responsibility for all AI-FRTD systems along with their health, modifications and upgrades will be assigned to the Department of Homeland Security.

TECHNOLOGY BACKGROUND

Artificial Intelligence Conceptual View

The idea of Artificial Intelligence has been around for decades, but even longer in existence, is the human desire to codify what thought is and the laws that apply to it. For example, in the 4th Century B.C. Aristotle proposed a deductive scheme for thought and logic in his syllogism "common-sense rules about how we think" (Moral Robots, n.d.).

In less ancient times, during the 17th Century, Thomas Hobbes, author of the Leviathan, was famous for professing the notion of ratiocination, which is the process of "a reasoned train of thought" (https://www.merriam-webster.com/dictionary/ratiocination). He felt "thinking consists of symbolic operations" and that thoughts are "special brain tokens, which Hobbes called phantasms or thought parcels." (History-Computer, n.d.) This idea of a packetized thought processes is a foreshadowing of what we now call Neurons.

In the 1950s, scientist began to study how human though process functioned. Alan Turing suggested the concept of a "thinking" machine that could interact with humans. In 1952, Sir Alan Hodgkin and Sir Andrew Huxley proposed what is referred to as the Hodgkin-Huxley Model, which is a "model of the brain as neurons forming an electrical network, with individual neurons firing all-or-nothing (on/off) pulses." (Foote,2016). Then in 1956 the first open use of the term Artificial Intelligence came from the Dartmouth Summer Research Project on Artificial Intelligence conference which was organized by Dr. John McCarthy. Professor McCarthy, a Professor of mathematics at Dartmouth College, proposed at the conference that "...every aspect of learning or any other feature of intelligence can in principle be so precisely described that a machine can be made to simulate it" (Kapp,2006).

During the Cold War of the 1960s and 1970s, investments were made in AI research with much of it coming from the Defense Research Projects

Agency (DARPA). (World-Information, n.d.). Unfortunately, the technology did not advance much during this period and it took until the 1980s for interest in AI to return with new efforts to develop and deliver AI, or "Expert Systems" as they were also known. (Foote,2016). In 1980, Digital Equipment Corporation (DEC) XCON system was an order entry system for the computer technology the company sold. XCON was designed to validate orders from DEC sales people using a rules-based approach to validate orders for missing or incorrect items. "At its peak, XCON had 2,500 rules...XCON was the first computer system to use AI techniques in solving real world problems" (Foote,2016). For a brief period in the late1980s until around the first part of 1993 interest in AI declined, led by the distraction of the rise of distributed computing known as the Personal Computer. However, as the 1990s progressed computer systems became more internetworked due to creation of the DARPANet that evolved into the NSFNet and finally the Internet.

Throughout the beginning of the 21st century, the distributed computing model has evolved into a centralized model, with large-scale computing centers housing an ever-increasing amount of data and processing power. These large-scale computing centers, or clouds as they are sometimes referred to, are managed by large organizations with dedicated teams supporting and selling their storage and compute capacity. The processing power contained in the computers that make up data centers has increased exponentially due to major advancements in microprocessor technology. In addition to increased computing power, there are now multiple options for microprocessor logic types that can be chosen based on the processing it will conduct. For example, the Graphic Processing Unit (GPU) Microprocessor which was originally designed to improve display graphics for visually intensive programs like Computer Generated Animation and Computer Games have been found to be superior to Central Processing Unit (CPU) Microprocessors for applications that are mathematically intensive such as Artificial Intelligence algorithms. (Wheeler, 2017)

Better than humanly possible, properly equipped computers can accept more input, store greater amounts of data, sort and recall it with greater speed and accuracy. Also, computers provide greater data resiliency with backup to other computers and long-term archive systems capable of storing data for decades. Since computers can input, output and manage data at a rate beyond the human capacity, complex data warehouse models have evolved to compliment human decision making. As a side effect of having the ability to store and process vast amounts of information, discovery of previously unknown patterns in data is occurring. This discovery of data patterns is important since it supports innovation and improves the quality of decision making made by humans.

Beyond the improved data collection, storage, retrieval and data analysis that computers can perform over humans, there is a desire to improve how a computer processes its input, with the intent to deliver more sophisticated output. To achieve this objective, computer science has been improving Artificial Intelligence in many areas such as autonomous vehicles, medical research, facilities management, data analytics, robotics, facial recognition, and even "natural-sounding synthetic speech from text like the Estonian language." (Fishel-Mihkla,n.d.)

Fundamentally computers are binary calculation devices that solve mathematical equations. A computer converts input it receives into binary data which is analyzed using mathematical algorithms. Artificial Intelligence is a general term that encompasses several ways to use computers to conduct specialized mathematical calculations that provide more sophisticated output. The objective of the various components of Artificial Intelligence is to improve the intrinsic value of the output it generates from input it receives. The data fed into a computer may be manually input by humans, input from other computing devices or retrieved by the system itself.

The Neuron

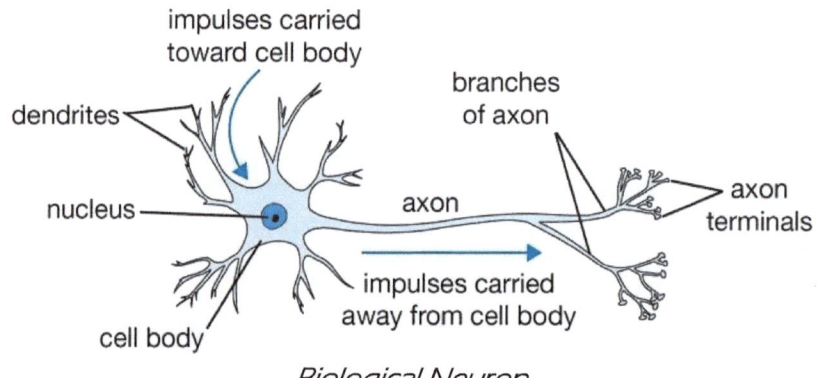

Biological Neuron
https://wiki.tum.de/display/lfdv/Artificial+Neural+Networks

"A nerve cell, or neuron, is a cell that receives information from other nerve cells or from the sensory organs and then projects that information to other nerve cells, while still other neurons project it back to the parts of the body that interact with the environment, such as the muscles". (National Research Council, p.116). More specifically, in a biological neuron, the dendrites of the neuron receive messages from other nerve cells or neurons, then transfer it to the nucleus of the neuron (also known as the soma) which processes the message then sends a response down the axon to other nerve cells' dendrites which in turn effects a reaction such as a muscle movement. The axon can also send it to the dendrites of another neuron for that cell's soma to process it further. The connection area between a sending axon and the receiving dendrite is known as the synapse. In biological entities, the synapse can be either an electrical or chemical connection between the two neurons. If a message is sent down the axon, to be forwarded out the neuron to another neuron, the synapse at the end of the axon, known as the axon terminals, will attempt to send either a chemical or electrical signal to another neuron's dendrite that also participates in the same synapse area.

Depending on the signal sent by the (presynaptic) neuron's axon terminal across the mutually shared synaptic region, the receiving neuron's (postsynaptic neuron) dendrite will either accept or reject the message.

In the world of Artificial Intelligence, the concept of a neuron exists, but how input is accepted is based on mathematical representation, not electrical or chemical stimuli as in a biological neuron.

After the Dartmouth AI conference Paul Rosenblatt-with inspiration from Warren McCulloch, Walter Pitts and Donald Hebb-formalized a conceptual perspective for an artificial neuron by defining a "perceptron" (CalState-LB, n.d.). The perceptron which Rosenblatt postulated in his book, Principles *of Neurodynamics,* was presented as the foundation for a visual recognition computer designed for the Department of Defense (DoD,1961).

Rosenblatt's perceptron theory provided a mathematical algorithm for a "brain model" (DoD, p.3) that accepts weighted input values that represents the priority of the input, adds a bias value with the resultant value(sum) determining the next action. For example, if the sum of the weighted inputs reaches a pre-defined threshold in the receiving neuron then the value will be accepted as an output for the perceptron, if not, the perceptron will ignore the inputs and not provide an output.

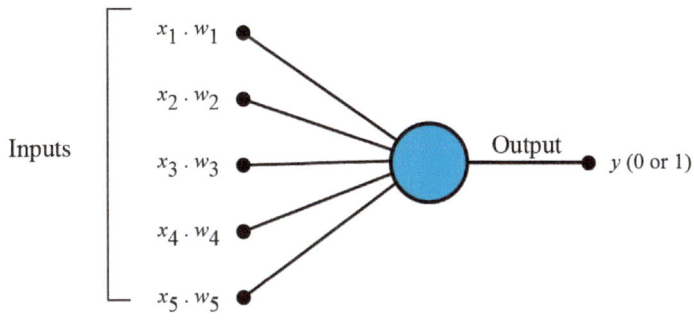

Artificial Neuron *(Perceptron)*
https://wiki.tum.de/display/lfdv/Artificial+Neural+Networks?preview=/23573754/23573796/neuron.png

Beyond Rosenblatt's original perceptron design, additional mathematical operations are applied to a perceptron to enable it to function in a neural network. For example, a Sigmoid function, tanh and a Rectified Linear Unit (ReLU) calculation is applied to the values. These calculations are applied to the input weighted values to normalize the summed values to create what is known as the "Activation Function", which is the process to determine if a perceptron should send (activate) an output. (Osserman, 2017)

Rosenblatt's work on the DoD funded Perceptron Computer was not successful and the understanding of a perceptron was slow in coming. But over time his work for the DoD to create the core of a machine using a perceptron was revolutionary enough to be the seed for the creation of Artificial Intelligence.

To clarify, a perceptron is not a neuron in a traditional sense since it works by mathematical stimulation and not chemical or electrical stimulation like a biological neuron, however, it has become customary to refer to a perceptron as a neuron.

Artificial Neural Network

Like a Biological Neural Network (BNN) that is comprised of multiple interconnected neurons, an Artificial Neural Network (ANN) is an interconnected network of more than one artificial neuron (perceptron). In both a BNN and ANN, inputs or outputs to an individual neuron may be received from another neuron or from a non-neuron (stimulus), but a neuron must have at least one connection (input or output) to another neuron to be a part of a neural network.

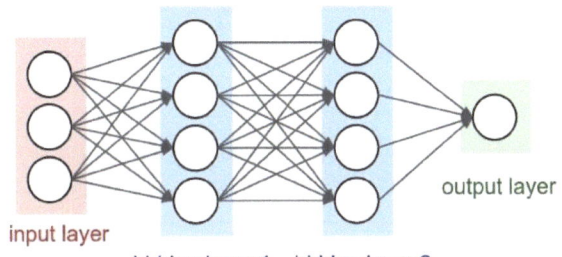

Image Courtesy of: http://cs231n.github.io/convolutional-networks/

In an ANN, if a neuron only accepts input from another neuron and outputs to another neuron, this is considered a "hidden layer". This distinction of a hidden layer is unique to ANN and does not apply to BNNs since "the human brain is not structured this way." (King, 2017), The impact of a hidden layer is important when it comes to understanding how an ANN must learn to improve the accuracy of its final output since there is a hidden layer that is not directly communicated with when providing feedback during training, thus techniques to modify the calculations done by the hidden neuron to improve its "intelligence" must be utilized.

Intelligence

The Oxford Dictionary defines "intelligence" as "The ability to acquire and apply knowledge and skills". To acquire knowledge and skills, an entity must learn. Therefore, learning is key to intelligence.

To derive an output, an artificial neuron accepts weighted mathematical values inputs from other neurons or external systems. The neuron processes these values using a form of linear algebra (Specifically Matrices) and the resultant sum is passed to another neuron or an external system, such as sensory neurons. If the output of a neuron is incorrect, a method must exist to modify the incorrect value.

To help derive the proper output, a neuron or ANN can be programed the response algorithm in advance (hard coded), which is fine for simple decision tasks. However, for more complex decisions, programming the path to determine a correct output may be too difficult. In fact, being able to eliminate the requirement to predefine an outcome is one of the key capabilities of Artificial Intelligence. If properly trained, an ANN can process larger amounts of information, more complex variables, all at a faster rate than a static computer program or a BNN.

When the output of a neuron is not programmed in advance, other models for learning can be utilized. The general models for learning are known as supervised learning, unsupervised learning, semi-supervised learning and Reinforcement Learning (RL). These models can be used for a neuron to learn the correct output (Rodriquez,2017)

Supervised learning is a method of learning "that uses a known dataset (called the training dataset) to make predictions". (MathWorks-Supervised Learning, n.d.) One supervised learning technique used to train an ANN is known as a Feed Forward Network (FFN). An FFN uses a known dataset, referred to as "Labeled Data", which is a paired set of data that contains an input value with its correct output value. The input data is fed into a Neural Network (NN) and the final output from the NN is compared to the correct value. If the output is incorrect, the weights(bias) of the input values will be adjusted. The complete process from initial input into a NN to the final output of the NN is referred to an "epoch of learning". (djmw,2004) The epoch will be repeated until the final output value is correct.

Another learning technique, known as Backward Propagation can also be used for supervised learning of a NN and a Deep Learning (DL) network. A DL network is essentially an ANN that is trained using very large datasets (MathWorks-Deep Learning, n.d.). Backward propagation is a process that attempts to send back (back propagate) recommend adjustment values to the neurons in the neural network (Robin,2009). The method works by repeating the epoch, comparing the actual output to the expected value, then using gradient descent and linear regression to modify the value to descend (close the gap) the actual output value towards the expected output value. (Nedrich, 2014)

There are two main approaches a supervised learning model uses for its datasets. The first type of use for a dataset is for use in a regression analysis model, such as a backward propagation or a deep learning network which are used for iterative learning. The second approach for using a dataset is called classification. In classification, there are groupings of similar data defined upfront and subsequent data is predicated relative to a predefined group. When a variation occurs from the expected result (i.e. predication inaccuracy) a value or weight is determined then applied to adjust the prediction in subsequent inputs. This approach is favored for the FFN technique.

Another supervised learning methodology is the Support Vector Machines (SVM). An SVM supervised learning method uses both clustering of multiple dimensional data points and regression analysis of the multi-dimensional dataset. The benefit of this two-group approach is greater accuracy and reduced training times. Instead of basing training only on single groupings (clustering) of similar data or iterative training to reduce variance (regression analysis), SVM looks to "formulate the problem in a difference space" (Phillips, p.804) by seeking optimal adherence to both clustering and regression analysis simultaneously.

Unsupervised learning-unlike supervised learning-does not use a predefined input or output dataset to compare values going in or coming out of the neurons. With unsupervised learning, data is fed (input) into the neuron or ANN and "Unsupervised learning algorithms group the data in an unlabeled data set based on the underlying hidden features in the data", (Jones,2017) It is this method of self-identifying patterns that has led to the discovery of previously unknown patterns in data when using AI to conduct the analysis of large datasets.

In security, unsupervised learning is the conceptual foundation for the creation of Generative Adversarial Networks (GAN). GANs have been used to create malicious software (Malware) for nefarious purposes and as a tool to protect against malicious software. These unsupervised learning networks are referred to as MalGANs. (Weiwei-Ying,2017)

Semi-supervised learning, as its name implies, is a hybrid or combination of learning methods where known datasets (labeled data) are used with unknown datasets (unlabeled data). "The goal of semi-supervised learning is to understand how combining labeled and unlabeled data may change the learning behavior, and to design algorithms that take advantage of such a combination."(Xiaojin-Goldberg,2009) This method of learning is more akin to how BNNs work, meaning humans are typically presented with information, they understand part of it from prior learning, but there may be a part of the information that requires them to learn how to respond to it. An example of information presented that is a combination of old and new knowledge happens when learning a new skill or traveling to an unfamiliar area.

Reinforcement Learning is "a goal-oriented learning based on interaction with [an] environment". (Shaikh,2017). The distinguishing characteristic RL has over supervised, unsupervised and semi-supervised learning is that a feedback or "reward function" (Shaikh,2017) exists that can be used to independently improve the values (weights) of the inputs.

As previously mentioned, computers can identify patterns and interrelationships in data previously unknown, especially when analyzing large datasets. Along with their ability to solve problems that require analysis of large datasets, the analysis they perform, with any size dataset, is done much faster than humanly capable. If an AI system can evaluate its inputs and outputs independently it can self-achieve accuracy. Interestingly, along with its self-improvement, it can offer output improvements to the system it is sending its data. For example, Google's AutoML AI attempts to first learn the required proper outputs for a given AI target, once it is successful, it will then seek to modify its AI target or create a "child" object with the experience it has gained during its RL phase (Galeon-Houser, 2017). An example of creating and improving an AI child is Google's AutoML improvement to its NASNet. NASNet is a Machine Learning (ML) based AI child to Google's AutoML.(Google Brain Team, 2017).

An example of an ML based AI modifying its output, to improve its quality was the result of an experiment conducted by Facebook researchers in the Facebook AI Research Lab (FAIR)with two chatbots(https://aws.amazon.com/what-is-a-chatbot/). The researcher's setup up two chatbots to talk to each other unrestricted. After a brief span of time of independent communications, the chatbots modified their outputs to improve how they communicated with each other. Interestingly, the communication protocol they developed may have been optimal for them and allowed them to communicate more efficiently, it was incomprehensible to humans. (Griffin, 2017)

Computer Vision and Facial Recognition

Facial Recognition Technology (FRT) has evolved to the point of being considered a form of Biometric identification that can uniquely identify a person.

In the 1960s, the effort began to develop a method for facial recognition. One technique developed at this time would compare images of faces and "calculated distances and ratios to a common reference point" (DuVal, n.d.) until a match was found. In the 1970s the team of Goldstein, Harmon and Lesk developed a technique of facial recognition that examined 21 specific facial features, such as lip thickness and hair color. (FBI-Documents, n.d.) However, the accuracy of these early additive techniques required manual effort to compare and measure images, in addition the accuracy rate was low and required clear face-forward images.

What began the successful march to improving and eventually automating facial recognition was pursuing a more subtractive and clustering approach. In the late 1980, Sirovich and Kirby proposed what they called an "eigenpictures" (Sirovich-Kirby, 1986) technique which is essential applying eigenvector analysis to images. So, rather than trying to compare facial images to one another and looking for a match, they proposed the inverse logic of looking for variance to a baseline. Then the research team of Turk and Pentland proposed a working process for the eigenpictures technique in their scholarly article titled "Face Recognition Using Eigenfaces" (Turk-Pentland, 1991) which proposed a similar theory built upon identifying variances to standard characteristics contained in large data sets. Later, they further improved the model by proposing a method that used captured still images from moving images(video).

From Sirovich, Kirby, Turk and Pentland's work, we now had a method that could be automated. This identification method, or Principal Component Analysis (PCA) as it is known, was tested at the 2001 Super Bowl to detect wanted criminals. It "helped identify 19 individuals in the crowd who had outstanding warrants" (Rogers, 2016). This enhanced surveillance was not well received by the public who expressed concerns over excessive police surveillance. Later that year, on September 11th, attitudes became more accepting of enhanced surveillance methods to protect against attackers.

Alex Pentland of the Turk-Pentland team was later one of the scientists awarded a grant from a research project managed by The National Institute of Standards and Technology (NIST) and Defense Advanced Research Project Agency (DARPA) for the Department of Defense Counterdrug Technology Development Program Office. The project started in 1993 with the intent to improve and automate what is known as Face Recognition Technology (FERET).

To achieve the objective of automated FERET, project members sought to improve the mathematical algorithms used for facial recognition. When the project was done, NIST stated: "The FERET Program was a highly successful effort that provided direction and credibility to the facial recognition community" (NIST-FERET, n.d.)

Beginning in 2000, the NIST began the first of their Face Recognition Vendor Test (FRVT). The focus of the testing was to examine new algorithms and their accuracy being developed by researchers in the commercial sector. Since 2000, another 6 FRVTs have occurred, with another one scheduled in 2018 (NIST-FRVT, 2017). The NIST also held a public challenge known as the Face Recognition Grand Challenge (FRGC) that ran between 2004 and 2006. The challenge was more focused on facial recognition improvements for the U.S. Government. As the challenge progressed, the datasets of images presented for analysis by the researches became increasingly complex. The knowledge gained because of the FRGC challenges have been instrumental in expanding both knowledge of facial recognition and to provide direction for future research. (NIST-FRGC, n.d.)

The paradigm shift in thinking PCA brought to facial recognition has been very important not only because of its improved recognition capability but it has fueled complimentary techniques like Linear Discriminant Analysis

(LDA) (Yang, et al. n.d.). Like PCA, LDA is based on variances of an image against defined parameters, however LDA has a greater discrimination capability so it delivers improved facial recognition accuracy.

The previous methods for facial recognition relied on linear analysis of a facial image. Despite this improved methodology, it still yields less accuracy in detection due to environmental factors such as lighting, atmospheric clarity, a subject's pose and their expression. A greatly improved method over the linear face analysis method uses a technique known as Elastic Bunch Graph Matching (EBGM)(Wiskott, et al., n.d.). EBGM uses a method of applying a special filter over a facial image. The filter is a texture mapping filter known as a Gabor Filter. (Murthy,2014) Using this method, "image analysis approaches based on Gabor filter conceptually imitate the human visual system (HVS)." (Um, et al., n.d.). Specific points on the grid are superimposed on the image, then analyzed for their uniqueness and become the identity of the image. Along with providing greater accuracy in facial recognition-since the filter is a subtractive filter-it reduces the image size and areas that must be examined so recognition time is reduced.

Convolutional Neural Network for Facial Recognition

A Convolutional Neural Network, like an Artificial Network, is composed of artificial neurons that have inputs and outputs and rely on mathematical calculations to determine output. However, in contrast to Artificial Neural Networks that are 2 Dimensional networks, a ConvNet has a third dimension which makes it ideal for image processing.

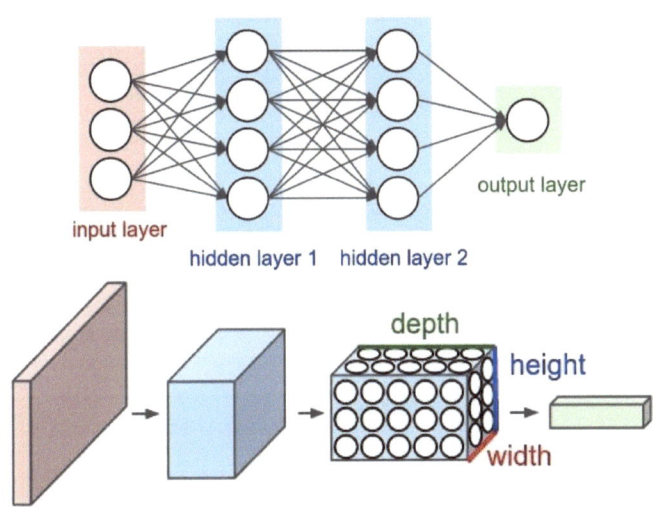

Left: A regular 3-layer Neural Network. Right: A ConvNet arranges its neurons in three dimensions (width, height, depth), as visualized in one of the layers. Every layer of a ConvNet transforms the 3D input volume to a 3D output volume of neuron activations. In this example, the red input layer holds the image, so its width and height would be the dimensions of the image, and the depth would be 3 (Red, Green, Blue channels).
Image Courtesy of: http://cs231n.github.io/convolutional-networks/

By adding a third dimension to a NN, it can greatly reduce the complexity required for a standard ANN to manage the input weights for all the neurons required to encode an image. For example, a small image of 32x32 pixels with only 3 color channels (Red, Green and Blue) would require 3,072 weights (32x32x3). Of course, this is a small image, a more workable image size would be at least 200x200 pixels, again with 3 color channels the neurons would have over 120,000 weights (200x200x3).

In addition to reducing complexity and size of a NN, by adding a third layer it becomes similar to how a BNN is constructed, which is very good at image processing. The discovery of how a biological brain conducts image processing came from the groundbreaking work done by David Hubel and Torsten Wiesel. In the 1950s, Hubel and Wiesel set out to map the visual cortex of a mammal by examining a cat's visual cortex. (Fehlhaber,2014) Hubel and Wiesel determined the neurons in a cat's visual cortex were arranged in a 3-Dimensional array. They discovered this when they ran experiments by placing recording electrodes in the visual neural receptors of a cat's brain. They presented lines to the cat's eyes, then monitored the audio responses(clicks) made when a visual neuron was stimulated. Throughout the following decades they continued to research and verify the visual cortex and its neural construct. In 1981, they were awarded the Nobel Prize in Physiology and Medicine for their work.

"Convolutional neural networks are the state of the art technique for image recognition" (Koehrsen, 2017). In 1998, Yann LeCun, Leon Bottou, Yoshua Bengio and Patrick Haffner published a paper that proposed an architecture for recognizing handwriting numbers and letters (LeCun, et al.,1998). They later called the architecture, LeNet 5.

Like the Gabor filter method that applies a filter to an image to identify and record unique details, a ConvNet supported image analysis applies a filer to identify the details of each pixel. This image analyses will be repeated for each depth (color) of the image. For example, if the image has a depth or 3, then three filters will be applied to the same image for each depth.

In summary, because of improvements brought on by PCA analysis, LDA analysis, the use of AI's Machine Learning Support Vector Machine (SVM) algorithms and the increased use of ConvNet for facial recognition, computers ability to accurately and rapidly recognize human faces is quickly improving.

THREAT ASSESSMENT

The Safe School Initiative (SSI) is a joint initiative between the U.S. Department of Education and U.S. Secret Service that began in 1999 as a response to the Columbine High School shooting. The SSI was established to help in "identifying pre-attack behaviors and communications that might be detectable-or "knowable"-and could help in preventing some future attacks" (SSI-Report, p. ii). In their final report, the SSI initiative offered numerous profiles of previous shooters, however no conclusive profile was offered, in fact they stated, "There is no accurate or useful "profile" of students who engaged in targeted school violence" (SSI-Report, p.11)

The Critical Incident Response Group (CIRG) of the National Center for the Analysis of Violent Crime (NCAVC) at the FBI Academy issued a report titled, "The School Shooter: A THREAT ASSESSSMENT PERSPECTIVE". Their report did not provide a conclusive profile of a school shooter either, but they did offer what they called "The Four-Pronged Assessment Model". Their model provides an "assessor" a roadmap of items they should collect on an individual when a threat is received. If the individual or "student appears to have serious problems in the majority of the four prongs...the threat should be taken more seriously and appropriate intervention by school authorities and/or law enforcement should be initiated as quickly as possible". (CIRG, p.10) If the one making the threat is a student, the assessor should be a staff member of the school the student attends. The recommendation from the report states: "The assessor may be a school psychologist, counselor, or other staff member or specialist who has been designated and trained for this task." (CIRG, p.11)

In the findings the SSI did report one early warning sign that occurred in a high percentage of the shootings examined, which the FBI-CIRG also called it out as a concern and referred to it as "Leakage". Leakage means "when a student intentionally or unintentionally reveals clues to feeling, thoughts, fantasies, attitudes, or intentions that may signal an impending violent act" (CIRG, p.16). From an analysis of 37 incidents of school shootings involving 41 attackers over a 26-year period running from 1974 to 2000 it was determined that 81 percent of the time "at least one person had information that the attacker was thinking about of planning the school attack". (SSI–Report, p.24)

SOCIAL MEDIA

Monitoring social media sites for leakage is becoming common place. Social media sites have published documentation on their Application Programming Interface (API) for programmers to write code to extract data from the social media sites. Data such as verbiage posed by users and their profile pictures along with others they have posted can be harvested. Programming skills are not required to retrieve data from social media sites, since there are plenty of free tools for social media monitoring that take minimal effort to learn to use. In addition, there are professional tools and services that will search for key words and ones that assess social media content for behavioral threat indicators (DHS-NCCIC,2014). The tools and services can proactively notify people responsible for the safety of a location by sending the name of the person of interest, their picture and the concerning statement(s) that were posted.

From their beginning, social media sites have provided capabilities to reach into their databases and harvest information on their users. It is this harvesting of millions and now billions of people's information that provides a very valuable resource for social analysis. For example, Information about the users regarding their 'likes', pictures of themselves, members of their family, who they consider their 'friend", where they like to travel, entertainment methods, and much more can be a valuable harvest for companies and people that use this information for their research purposes. To make it easy to retrieve information social media sites created APIs and published their format and documented how to integrate them for those wishing to use other applications to retrieve the information. If a company or individual does not want to undertake the development effort to build a system that can retrieve from the social media sites they can acquire various social media extraction tools that are already configured to retrieve data from social media sites. Some of the harvesting tools cost money, while others are free (Mindruta, 2017). Once the user has the software, generally all they must do is request the proper account from the social media site and put it into the software.

If a company would rather rely on the expertise of companies whose sole business it is to conduct social media analysis they have many options, in fact "As social media has become a mainstream business tool, companies continue to invest into social media programs such as social media monitoring..." (Ideya, 2017). Social Media Monitoring (SMM) has become a mature industry with many companies collecting, analyzing and presenting various analytics (*Appendix F*) for sale to customers. The analysis of social media participants for various purposes can be found, ranging from consumer targeted sales and marketing to government-oriented research companies such as political campaign analysis (Cambridge Analytica, 2018) and Social Media Intelligence (SOCMINT) gathering.

There are even formal training and certification for social media analysis, such as the Certified Social Media Intelligence Expert (CSMIE) offered by the National Initiative for Cybersecurity Careers and Studies (NICCS) (NICCS, n.d.).

Select school districts in the U.S like ones in Massachusetts and Miami-Dade County (Fussell, 2018) are either already using or preparing to use services, software or training that are specifically created to examine social media sites to detect leakage for an early warning of a potential student attacker.

FEATURE ADD-ONS

This paper presented one solution for an Artificial Intelligence monitored environment. Specifically, that of facial identity matching to selected federal and social media databases to identify and alert of potential threating people entering or located within a monitored site.

It should be noted that this same AI based system could be expanded to provide greater protection. For example, the system can use subtractive analysis to conduct area monitoring for anomalies that could present risk, such as changes to areas by threat actors that an AI based camera monitoring system can easily detect but may be difficult for a human to do the same. For example, changes to an area that provide physical advantage points (An above ground level, open window that can be used by a sniper), unusual human surveillance (Such as a person(s) taking unusual number of pictures or notes of an area), anomalous vehicles (Such as rented moving trucks parked outside of a federal building) or unusual activity in a high-risk area (such as intersections that are accident prone).

In addition to anomaly detection using subtractive analysis, this same system with slight additions could detect weapons and body armor on those at a site and send an alert using the same Alerting System to those charged with protecting a site.

The solution proposed will continue to lower in cost and improve in sophistication as manufacturers implement in their products the improvements in facial recognition technology proposed by standards bodies, such as ISO/IEC 19754-5:2011(ISO, n.d.) and the INCITS 385-2004 standard already adopted by the DHS(INCITS,2004).

ACRONYMS

AI=Artificial Intelligence
AI-FRTD=Artificial Intelligence-Facial Recognition Threat Detection
ANN=Artificial Neural Network (Computer 'Brain')
API=Application Programming Interface
BNN=Biological Neural Network (Human Brain)
CIRG=Critical Incident Response Group
CNN=Convolutional Neural Network
ConvNet=Convolutional Neural Network
CPU=Central Processing Unit
DARPA=Defense Advanced Research Projects Agency
DL=Deep Learning
EBGM=Elastic Bunch Graph Matching
FERET=Face Recognition Technology (NIST Terminology)
FFN=Feed Forward Network
FRT=Facial Recognition Technology
FRTD=Facial Recognition Threat Detection
FRVT=Face Recognition Vendor Test
GAN=Generative Adversarial Network
GPS=Global Positioning System
GPU=Graphics Processing Unit
LAN=Local Area Network
MalGAN=Malicious Generative Adversarial Network
ML=Machine Learning
NCAVC=National Center for the Analysis of Violent Crime
NCCIC=National Cybersecurity and Communications Center
NCIC=National Crime Information Center
NIST=National Institute of Standards and Technology
NN=Neural Network
PCA=Principal Component Analysis
RBAC=Role Based Access Controls
RL=Reinforcement Learning
SOCMINT=Social Media Intelligence
SSI=Safe School Initiative
SVM=Support Vector Machines

APPENDIX

APPENDIX A: 5 Deadliest Shootings in United States History

APPENDIX B: Active Shooting Locations (2000-2016)

APPENDIX C: Active Shooting Incidents Per Year (2000-2016)

APPENDIX D: Active Shooter Casualty Breakdown Per Year (2000-2016)

APPENDIX E: Artificial Intelligence Facial Recognition Threat Detection Architecture Diagram

APPENDIX F: Mapping of Featured Social Media tools by type of social technology they offer

APPENDIX A: 5 Deadliest Shootings in United States History

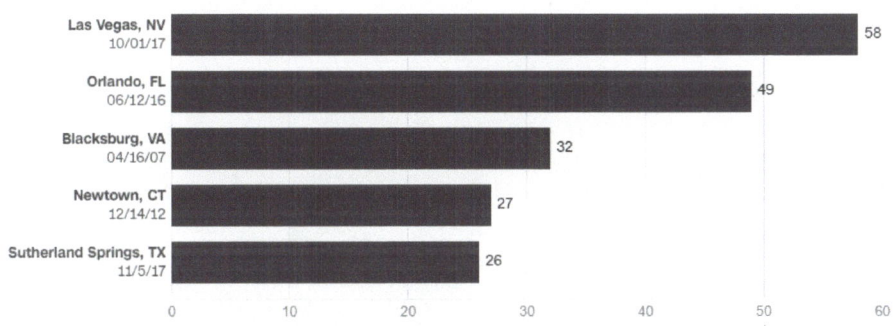

Willingham, A.J., Ahmed,S. (2017, November 6). Mass shootings in America are a serious problem – and these 9 charts show just why. 5 deadliest shootings. CNN. Retrieved from:
https://www.cnn.com/2016/06/13/health/mass-shootings-in-america-in-charts-and-graphs-trnd/index.html

APPENDIX B: Active Shooting Locations (2000-2016)

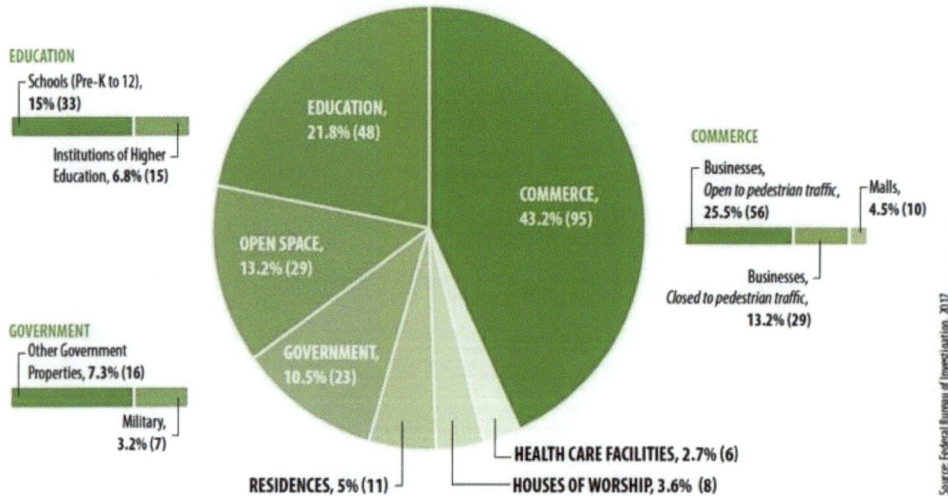

FBI (n.d.). Quick Look: 220 Active Shooter Incidents in the United States Between 2000-2016. Location Categories. Federal Bureau of Investigation. Retrieved from:
https://www.fbi.gov/about/partnerships/office-of-partner-engagement/active-shooter-incidents-graphics

APPENDIX C: Active Shooting Incidents Per Year (2000-2016)

Quick Look: 220 Active Shooter Incidents in the United States Between 2000 - 2016
Incidents Per Year

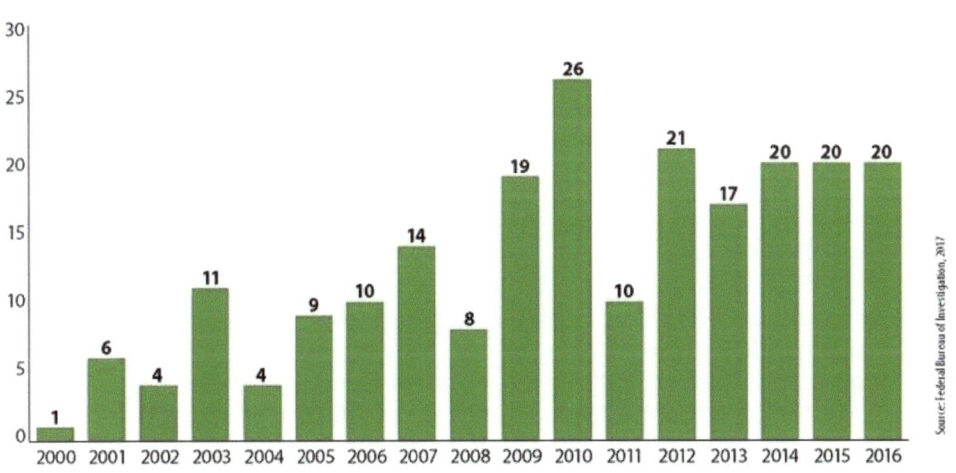

FBI (n.d.). Quick Look: 220 Active Shooter Incidents in the United States Between 2000-2016. Incidents Per Year. Federal Bureau of Investigation. Retrieved from: https://www.fbi.gov/about/partnerships/office-of-partner-engagement/active-shooter-incidents-graphics

APPENDIX D: Active Shooter Casualty Breakdown Per Year (2000-2016)

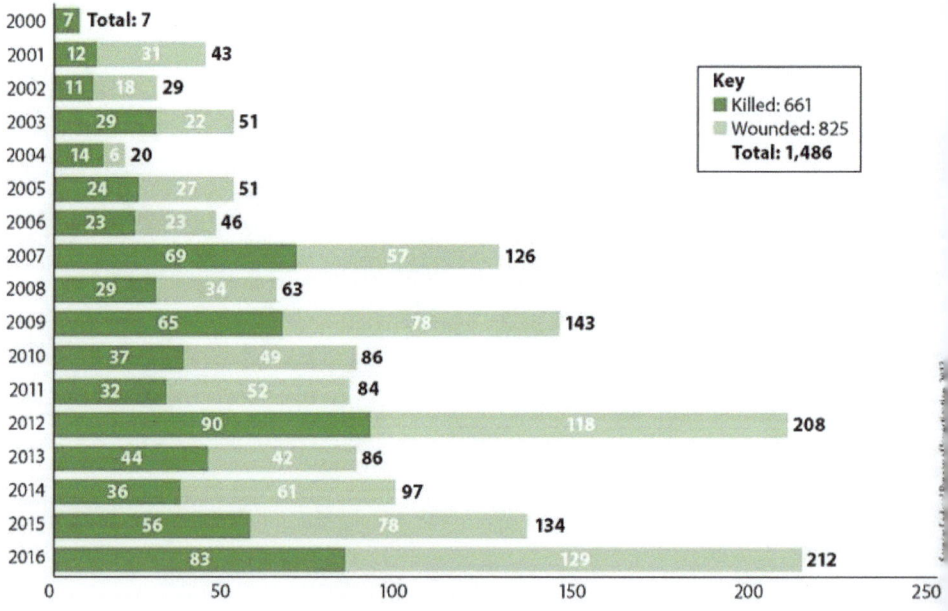

Quick Look: 220 Active Shooter Incidents in the United States Between 2000 - 2016
Casualty Breakdown Per Year

FBI (n.d.). Quick Look: 220 Active Shooter Incidents in the United States Between 2000-2016. Casualty Breakdown Per Year. Federal Bureau of Investigation. Retrieved from:
https://www.fbi.gov/about/partnerships/office-of-partner-engagement/active-shooter-incidents-graphics

APPENDIX E: Artificial Intelligence Facial Recognition Threat Detection Architecture Diagram

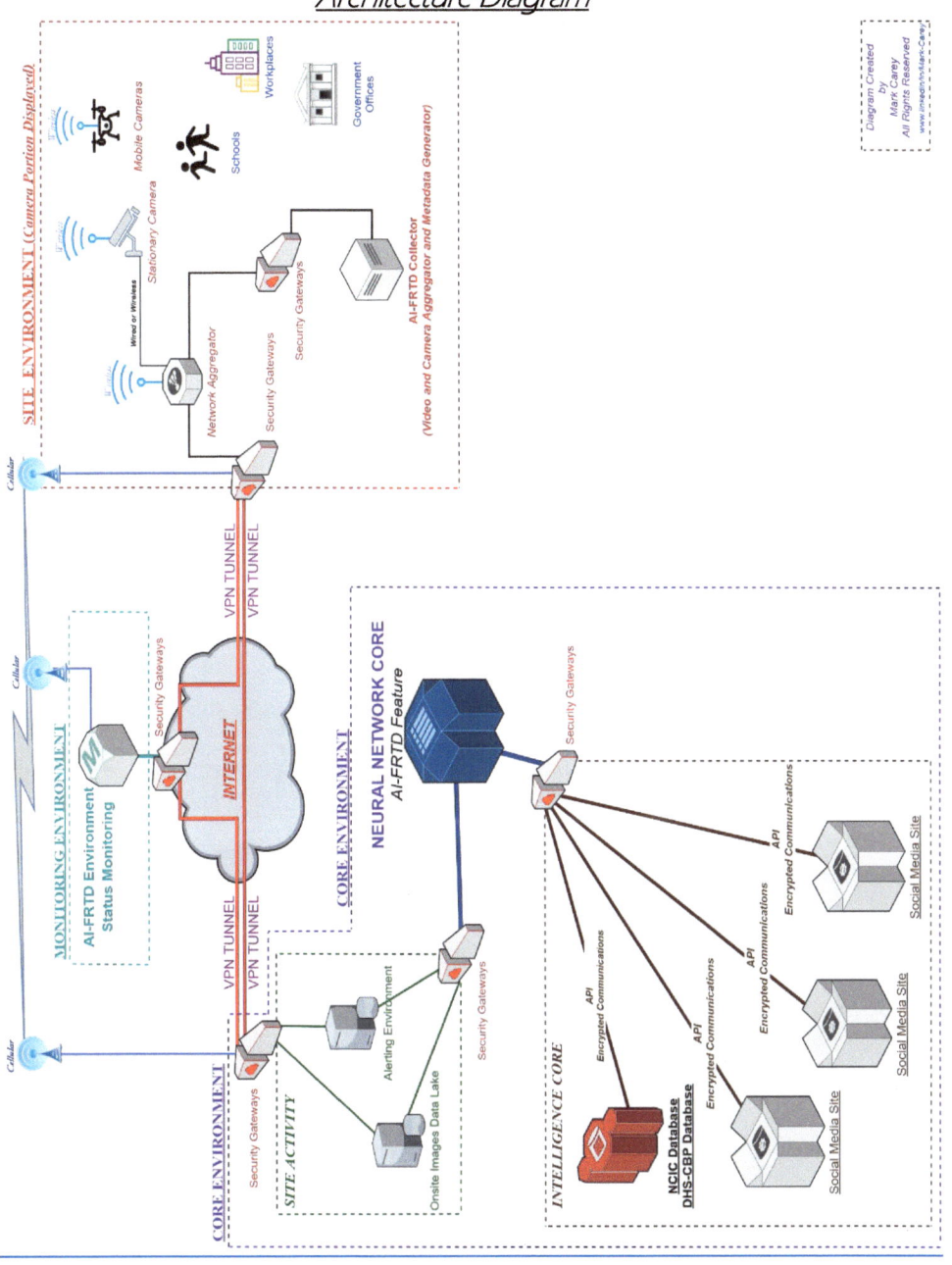

APPENDIX F: Mapping of Featured Social Media tools by type of social technology they offer

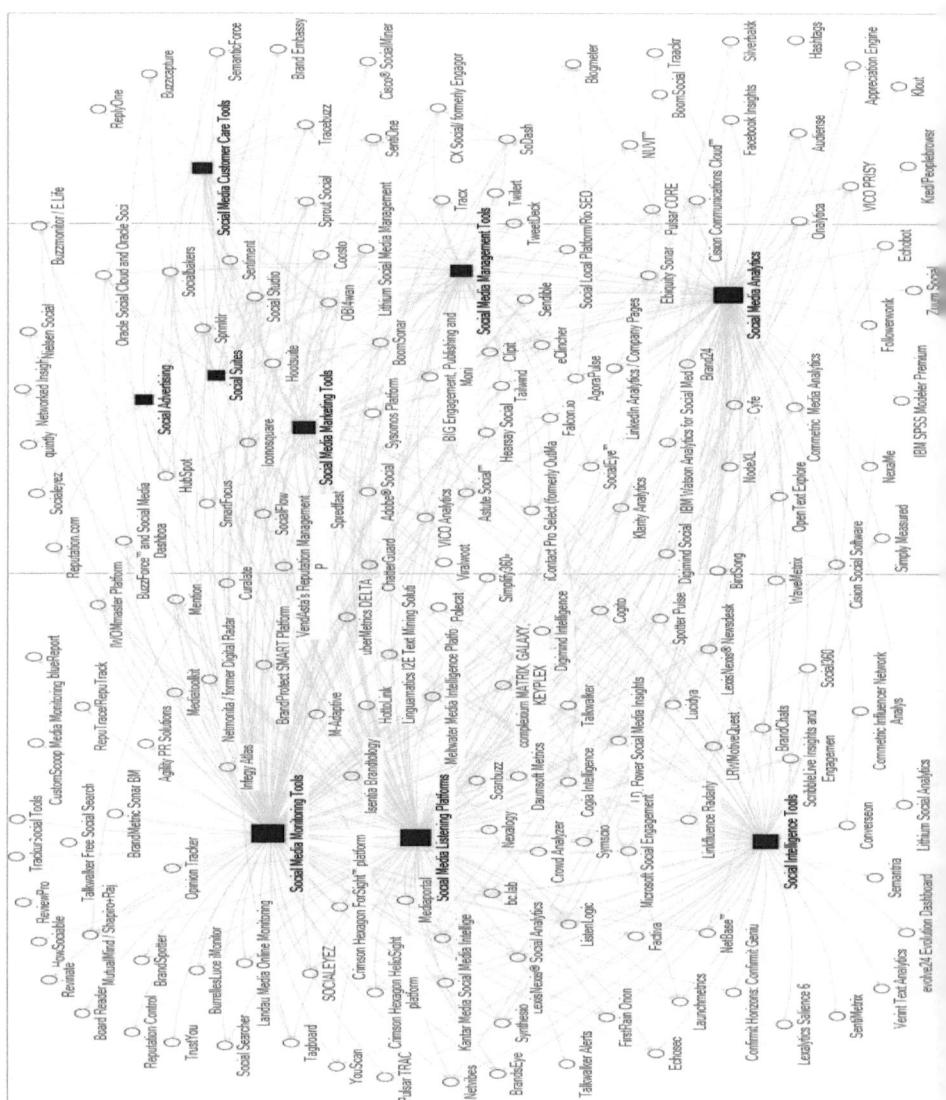

Image courtesy of:
http://ideya.eu.com/images/Social%20technology%20landscape%20map%20large.png

BIBLIOGRAPHY

1. Ahmed, S. (2018, May 18). There has been, on the average, 1 school shooting every week this year. CNN. Retrieved from: https://www.cnn.com/2018/03/02/us/school-shootings-2018-list-trnd/index.html
2. Coaston,J.,Kirby,J.,Stewart,E. (2018, May 20). Santa Fe, Texas, school shooting: what we know. Vox. Retrieved from: https://www.vox.com/latest-news/2018/5/18/17368598/school-shooting-santa-fe-texas
3. Cornell Law School (n.d.). Second Amendment. Legal Information Institute. Retrieved from: https://www.law.cornell.edu/wex/second_amendment
4. Knoll,J.L.,Annas,G.D. (2016). Mass Shootings and Mental Illness. American Psychiatric Association Publishing. Retrieved from: https://psychiatryonline.org/doi/pdf/10.5555/appi.books.9781615371099
5. Crenshaw, M. (2017, October 11). When are mass shootings acts of terrorism? Stanford Freeman Spogli Institute. Retrieved from: https://medium.com/freeman-spogli-institute-for-international-studies/when-are-mass-shootings-acts-of-terrorism-2bf08070b343
6. Williams, J. (2017, October 2). White American men are a bigger domestic terrorist threat than Muslim foreigners. Vox. Retrieved from: https://www.vox.com/world/2017/10/2/16396612/las-vegas-mass-shooting-terrorism-islam
7. FindLaw (2018). Fingerprints: The First ID. Retrieved from: http://criminal.findlaw.com/criminal-procedure/fingerprints-the-first-id.html
8. Recap (n.d.). What Makes A Data Center "Hyperscale"? Retrieved from: https://recap-project.eu/news/hyperscale-data-center/
9. FBI-NCIC (n.d.). National Crime Information Center (NCIC). Federal Bureau of Investigation. Retrieved from: https://www.fbi.gov/services/cjis/ncic

10. DHS-CBP (2017, June). Biometric Exit Process. Department of Homeland Security, U.S. Customs and Border Protection. Retrieved from: https://www.cbp.gov/sites/default/files/assets/documents/2017-Oct/biometric-exit-process-fact-sheet.pdf
11. DOJ (n.d.). State Sponsors of Terrorism. U.S. Department of Justice. Retrieved from: https://www.state.gov/j/ct/list/c14151.htm
12. DHS-GETS (2017, August 31). Government Emergency Telecommunications Service (GETS). Department of Homeland Security. Retrieved from: https://www.dhs.gov/government-emergency-telecommunications-service-gets
13. Ungerleider,N (2013, April 17). Why Your Phone Doesn't Work During Disasters-And How To Fix It. Fast Company. Retrieved from: https://www.fastcompany.com/3008458/why-your-phone-doesnt-work-during-disasters-and-how-fix-it
14. Moral Robots (n.d.). Making sense of robot ethics. Retrieved from: https://moral-robots.com/philosophy/aristotles-ai/
15. History-Computer (n.d.). Thomas Hobbes. Retrieved from: http://history-computer.com/Dreamers/Hobbes.html
16. Foote, K.D. (2016, April 5). A Brief History of Artificial Intelligence. Dataversity. Retrieved from: http://www.dataversity.net/brief-history-artificial-intelligence/
17. Kapp, S. (2006, July 24). Artificial Intelligence: Past, Present, and Future. VOX of Dartmouth. Retrieved from: http://www.dartmouth.edu/~vox/0607/0724/ai50.html
18. World-Information (n.d.).1960s-1970s: Increased Research in Artificial Intelligence (AI). Retrieved from: http://world-information.org/wio/infostructure/100437611663/100438659474

19. Wheeler, A. (2017, August 25). NVIDIA's Artificial Intelligence Boom: What Makes AI and GPU so Compatible? Engineering.com. Retrieved from: https://www.engineering.com/Hardware/ArticleID/15471/NVIDIAs-Artificial-Intelligence-Boom-What-Makes-AI-and-GPUs-so-Compatible.aspx
20. Fishel,M.,Mihkla,M.(n.d.). Modelling the temporal structure of newsreaders' speech on neural networks for Estonian text-to-speech synthesis. Institute of Estonian Language, University of Tartu. Retrieved from: https://pdfs.semanticscholar.org/8cb9/d38ab791fe5608da8e9b0a00393185ab24c3.pdf
21. National Research Council (2000). How People Learn: Brain, Mind, Experience, and School: Expanded Edition. Washington DC: The National Academies Press. https://doi.org/10.17226/9853
22. CalState-LB (n.d.). History of the Perceptron. California State University, Long Beach. Retrieved from: https://web.csulb.edu/~cwallis/artificialn/History.htm
23. DoD (1961, March 15). PRINCIPLES OF NEURODYNAMICS. PERCEPTRONS AND THE THEORY OF BRAIN MECHANISMS. Frank Rosenblatt, Director, Cognitive Systems Research Program, Cornell University for the Department of Defense, Contract Monr-2381(00). UNCLASSIFIED. Retrieved from: http://www.dtic.mil/dtic/tr/fulltext/u2/256582.pdf
24. Osserman,H.(2017, June 27). Aspects of Deep Learning: Activation Functions. x.ai. Retrieved from: https://x.ai/blog/aspects-of-deep-learning-activation-functions/
25. King, P. (2017, May 21). How many hidden layers are in the neural network of the human brain? Quora. Retrieved from: https://www.quora.com/How-many-hidden-layers-are-in-the-neural-network-of-the-human-brain

26. Rodriquez, J. (2017, January 19). Types of Artificial Intelligence Learning Models. Medium. Retrieved from: https://medium.com/@jrodthoughts/types-of-artificial-intelligence-learning-models-814e46eca30e
27. MathWorks-Supervised Learning. (n.d.). Supervised Learning. Retrieved from: https://www.mathworks.com/discovery/supervised-learning.html
28. djmw (2004, April 28). Feedforward neural networks 1.1. The learning phase. University of Amsterdam. Retrieved from: http://www.fon.hum.uva.nl/praat/manual/Feedforward_neural_networks_1_1__The_learning_phase.html
29. MathWorks-Deep Learning (n.d.). What is Deep Learning? 3 things you need to know. Retrieved from: https://www.mathworks.com/discovery/deep-learning.html
30. Robin (2009, November 26). BACKPROPOGATION. Artificial Intelligence. Articles on Artificial Intelligence. Retrieved from: http://intelligence.worldofcomputing.net/machine-learning/learning-by-back-propagation.html#.WtoUuYjwZhE
31. Nedrich, M. (2014, June 24). An Introduction to Gradient Descent and Linear Regression. Atomic Object. Retrieved from: https://spin.atomicobject.com/2014/06/24/gradient-descent-linear-regression/
32. Phillips, P.J. (n.d.). Support Vector Machines Applied to Face Recognition. National Institute of Standards and Technology. Retrieved from: https://papers.nips.cc/paper/1609-support-vector-machines-applied-to-face-recognition.pdf
33. Jones, M.T. (2017, December 4). Unsupervised learning for data classification. IBM. Retrieved from: https://www.ibm.com/developerworks/library/cc-unsupervised-learning-data-classification/index.html

34. Weiwei,H. Ying,T. (2017, February 20). Generating Adversarial Malware Examples for Black-Box Attacks Based on GAN. Department of Machine Intelligence, Peking University. Retrieved from: http://www.cil.pku.edu.cn/publications/papers/2017/MalGAN_IJCAI_2017_Hu_Tan.pdf
35. Xiaojin,Z., Goldberg,A.B.(2009). Introduction to Semi-Supervised Learning. Synthesis Lectures on Artificial Intelligence and Machine Learning. Retrieved from: https://www.morganclaypool.com/doi/abs/10.2200/S00196ED1V01Y200906AIM006?journalCode=aim
36. Shaikh,F.(2017, January 19). Simple Beginner's guide to Reinforcement Learning & its implementation. Analytics Vidhya. Retrieved from: https://www.analyticsvidhya.com/blog/2017/01/introduction-to-reinforcement-learning-implementation/
37. Galeon, D., Houser, K.. (2017, December 1). Google's Artificial Intelligence Built and AI That Outperforms Any Made by Humans. Futurism. https://futurism.com/google-artificial-intelligence-built-ai/
38. Google Brain Team (2017, November 2). AutoML for large scale image classification and object detection. Retrieved from: https://research.googleblog.com/2017/11/automl-for-large-scale-image.html
39. Griffin, A. (2017, July 31). Facebook's Artificial Intelligence Robots Shut Down After They Start Talking to Each Other In Their Own Language. Independent. Retrieved from: https://www.independent.co.uk/life-style/gadgets-and-tech/news/facebook-artificial-intelligence-ai-chatbot-new-language-research-openai-google-a7869706.html

40. DuVal, A. (n.d.). Face Recognition Software. History of Forensic Psychology. Retrieved from: http://forensicpsych.umwblogs.org/
41. FBI-Documents (n.d.). Face Recognition. Federal Bureau of Investigation. Retrieved from: https://www.fbi.gov/file-repository/about-us-cjis-fingerprints_biometrics-biometric-center-of-excellences-face-recognition.pdf/view
42. Sirovich,L., Kirby,M. (1986, August 25). Low-dimensional procedure for the characterization of human faces. Division of Applied Mathematics, Brown University. Retrieved from: http://engr.case.edu/merat_francis/EECS%20490%20F04/References/Face%20Recognition/LD%20Face%20analysis.pdf
43. Turk,M.A., Pentland,A.P. (1991). Face Recognition Using Eigenfaces. Massachusetts Institute of Technology. Retrieved from: https://pdfs.semanticscholar.org/d5b2/061d88e09a50053e4027bc950cb981443ad8.pdf
44. Rogers,K. (2016, February 7). That Time the Super Bowl Secretly Used Facial Recognition Software on Fans. In pre-9/11, pre-Snowden America it was still a far-fetched idea. Motherboard. Retrieved from: https://motherboard.vice.com/en_us/article/kb78de/that-time-the-super-bowl-secretly-used-facial-recognition-software-on-fans
45. NIST-FERET (n.d.). Face Recognition Technology (FERET). National Institute of Standards and Technology. Retrieved from: https://www.nist.gov/programs-projects/face-recognition-technology-feret
46. NIST-FRVT (2017, July 8). Face Recognition Vendor Test (FRVT). National Institute of Standards and Technology. Retrieved from: https://www.nist.gov/programs-projects/face-recognition-vendor-test-frvt
47. NIST-FRGC (n.d.). Overview of Face Recognition Grand Challenge. National Institute of Standards and Technology. Retrieved from: https://ws680.nist.gov/publication/get_pdf.cfm?pub_id=150264

48. Yang,J.,Yu,H., Kunz,W. (n.d.). An Efficient LDA Algorithm for Face Recognition. Carnegie Mellon, School of Computer Science. Retrieved from: https://pdfs.semanticscholar.org/13b8/e93e463286a4b0275ecbb599569bbb7bf70f.pdf
49. Wiskott,L.,Fellous,J.M,.Kruger,N.,von der Malsburg,C. (n.d.). Recognition by Elastic Bunch Graph Matching. Institute for Neural Computation, Ruhr-University Bochum, Germany. Retrieved from: http://www.face-rec.org/algorithms/ebgm/wisfelkrue99-facerecognition-jainbook.pdf
50. Murthy,K.(2014, April 27). Computer Vision Tutorials. GABOR FILTERS: A PRACTICAL OVERVIEW. Retrieved from: https://cvtuts.wordpress.com/2014/04/27/gabor-filters-a-practical-overview/
51. Um,S.,Kim,J.,Min,D. (n.d.). Fast 2-D Complex Gabor Filter with Kernel Decomposition. Retrieved from: https://arxiv.org/pdf/1704.05231.pdf
52. Fehlhaber,K.(2014, October 29). Hubel and Wiesel $ the Neural Basis of Visual Perception. Knowing Neurons. Retrieved from: https://knowingneurons.com/2014/10/29/hubel-and-wiesel-the-neural-basis-of-visual-perception/
53. Koehrsen,W. (2017, July 27). Object Recognition with Google's Convolutional Neural Networks". Medium. Retrieved from: https://medium.com/@williamkoehrsen/object-recognition-with-googles-convolutional-neural-networks-2fe65657ff90
54. LeCun,Y.,Bottou,L.,Bengio,Y.,Haffner,P. (1998). Gradient-Based Learning Applied to Document Recognition. Retrieved from: http://yann.lecun.com/exdb/publis/pdf/lecun-98.pdf

55. SSI-Report (2004, July). The Final Report and Findings of the Safe School Initiative: Implications for the Prevention of School Attacks in the United States. United States Secret Service and United States Department of Education. Retrieved from: https://www2.ed.gov/admins/lead/safety/preventingattacksreport.pdf
56. CIRG (n.d.). The School Shooter: A THREAT ASSESSMENT PERSEPCTIVE. Critical Incident Response Group, National Center for the Analysis of Violent Crime, FBI Academy. Retrieved from: https://www.fbi.gov/file-repository/stats-services-publications-school-shooter-school-shooter/view
57. DHS-NCCIC (2014, May 2). Combating the Insider Threat. Department of Homeland Security, National Cybersecurity and Communications Integration Center. Retrieved from: https://www.us-cert.gov/sites/default/files/publications/Combating%20the%20Insider%20Threat.pdf
58. Mindruta,R. (2017, October 25). Marketing: Top 15 Free Social Media Monitoring Tools. Brandwatch. Retrieved from: https://www.brandwatch.com/blog/top-10-free-social-media-monitoring-tools/
59. Ideya (2017, November 9). Social Media Monitoring Tools and Services Report Public Excerpts 2017. Ideya Market Report, 8th Edition. Retrieved from: http://ideya.eu.com/images/Social%20Media%20Monitoring%20Tools%20and%20Services%20Report%20Public%20Excerpts%202017.pdf
60. Cambridge Analytica (2018). Data-driven campaigns. We find your voters and move them to action. Retrieved from: https://ca-political.com/?__hstc=163013475.6fa947136e32a82dbd135f79d433bcc0.1524182426372.1524182426372.1524182426372.1&__hssc=163013475.1.1524182426373&__hsfp=908707084
61. NICCS (n.d.). Certified Social Media Intelligence Expert(CSMIE). Retrieved from: https://niccs.us-cert.gov/training/search/mcafee-institute/certified-social-media-intelligence-expert-csmie

62. Fussell,S. (2018, March 24). US Schools Are Using AI To Check Students' Social Media For Warning Signs of Violence. Gizmodo. Retrieved from: https://www.gizmodo.com.au/2018/03/us-schools-are-using-ai-to-check-students-social-media-for-warning-signs-of-violence/
63. ISO (n.d.). ISO/IEC 19754-5:2011. Information technology—Biometric data interchange formats—Part 5: Face image data. International Organization for Standards. Retrieved from: https://www.iso.org/standard/50867.html
64. INCITS (2004) US. Department of Homeland Security Adopts INCITS Biometric Standard INCITS 385-2004. InterNational Committee for Information Technology Standards. Retrieved from: http://www.incits.org/news-events/press-releases/us-department-of-homeland-security-adopts-incits-biometric-standard-incits-3852004

www.ingramcontent.com/pod-product-compliance
Lightning Source LLC
Chambersburg PA
CBHW040240220526
45473CB00001B/313